Fabio R. de Araujo

Printed in Scotts Valley, CA - USA

MOTHER SHIPTON:

SECRETS, LIES, AND PROPHECIES

MOTHER SHIPTON:

SECRETS, LIES, AND PROPHECIES

Written by Fabio R. de Araujo, Historian

Text edited by June Fritchman

2009

Araujo, Fabio R., 1967-

Mother Shipton: Secrets, Lies, and Prophecies / Fabio R. Araujo – 1st ed.

ISBN: 978-85-62022-00-5

1. History. 2. Prophecies. 3. Legends.

TABLE OF

CONTENTS

17ᵗʰ century image depicting Mother Shipton as an astrologer revealing the future

INTRODUCTION

Secrets, lies, and prophecies. That will be one's experience if the subject is Mother Shipton. But who is Mother Shipton? Some will consider her the most famous prophet in the last 1000 years, after Nostradamus. In fact, she might be the most famous prophetess in the world's history. If some think she really existed and was a soothsayer or maybe an astrologer who predicted facts (most of them regarding the 17th century or about our own future), others may believe that she is no more than a legend slowly built over the years.

Different prophecies attributed to her by a few writers and biographical aspects have been published though the centuries. Meanwhile, from the end of the 19th century, some books have shown that lies and secrets have built what might be called a myth about a prophetess who perhaps never existed. Considering that we are unable to determine with certainty all aspects regarding her supposed life and existence, no matter how big our efforts may be, some secrets will probably remain uncovered forever.

In the following pages, this book will try to expose what seems to be the truth regarding her life and predictions. Although without doubt she reached an unparallel level as the most famous prophetess ever, was she a real prophetess? And, did she even indeed exist? Herein we will critically evaluate some old chapbooks about Mother Shipton, her biographical information based on an analysis done in the 19th century, and her last prophecies as attributed to her. We will try to analyze and understand how a local amusing myth, accompanied with some curious folklore, was created and developed by different people through the centuries, became the global myth it is today. We will also explore who might have been interested in spreading Shipton's tale, and what developments might have enhanced her fame and prophecies through generations.

CHAPTER 1

Mother Shipton is the most widely known and esteemed prophetess of the British Isles. Born in England about five centuries ago, this mysterious woman was the subject of many books printed in the 20th century that mention her as a certain Ursula Shipton. Throughout that century many questions arose, among them whether she really existed and about the validity of her prophecies. After all, what were her prophecies: Did she predict events that occurred in our modern times or only "see" events that had already taken place? Was her mother, Agatha Shipton, really a famous witch as some writers claimed? Was she really the "Devil's child" or is this just a legend?

This work will try to understand who Mother Shipton really was. It will highlight the issues of authenticity and attribution in her prophecies. It will also examine the supposed veracity of her biographical facts. In combining these efforts, it will make a distinction between what she *might* have said and what she certainly did not say. In any case, it is undeniable that Mother Shipton and her prophecies are part of history today. The fact some Encyclopedias mention her supports this statement. Britannica presents her as a

> witch and prophetess who is supposed to have lived in early Tudor times. There is no trustworthy evidence as to her ever having existed, but tradition has. Her maiden-name was Ursula Southill, Sowthiel or Southiel, and her parents were peasants, living near the Dropping Well, also called Petrifying Well, Knaresborough, Yorkshire. According to the legend, the cascading waters of the Well turned objects into stone. The date of her birth is uncertain, but it is placed about 1486-1488. Her mother, Agatha Southill, was a reputed witch, and Ursula from her infancy was regarded by the neighbors as 'The Devil's child'.

According to this legend, throughout her childhood, many pieces of furniture moved and flew around the house or inside the rooms. Columbia Encyclopedia describes her as a

> legendary English prophetess. She was first mentioned in an anonymous pamphlet, published in 1641[1], which described her

---

[1] This pamplet is published here, in this book; please see Chapter 9.

as having prophesied various events of the reign of Henry VIII and later.

Mother Shipton is mentioned in other encyclopedias and in many prophecy books from different countries, some of which are referred to later.

Legend or not, Mother Shipton acquired a life of its own and she is definitely still important in England. Some consider the Petrifying Well in Yorkshire the oldest attraction in England, and tourists from all over the world still visit it today. Another place visitors look for is the cave where she supposedly was born, known today as Mother Shipton's Cave. The cave, the Petrifying Well, and other attractions are run by Mother Shiptons Cave Ltd. Some locals still tell visitors she really existed and predicted the future, though this may be not so important anymore. A moth with "eyes" in its wings, the "Mother Shipton Moth," received her name a long time ago, and symbolizes she is part of the living world.

The only truth we can ascertain is that she became a source of legends, folklore, lies, prophecies, and secrets. If it is the case is that Mother Shipton is no more than a well-built perversion of facts, which today is not easy to know, it will always be a wonderful example of how myths are built by people's imagination throughout and how they survive and carry confusion to thousands, or in this specific case, maybe millions.

CHAPTER 2

The different aspects concerning the Mother Shipton's life, as published in various almanacs, books, and chapbooks during the previous centuries are often based on an old book written by Richard Head, the earliest extant edition of which is dated 1684. A small detail most subsequent writers did not know is that Head admitted afterwards that he forged most of the bibliographical details. This means most history about her published in the 20th century was fraudulent. According to this 1684 chapbook, Mother Shipton's mother, called Agatha, made love to the demon disguised as a handsome man. The book also says a "winged dragon" carried her once. Other books assert Agatha was a whore, so her mother's historical identity is disputed even in legend.

About 40 years previously, in 1641, (i.e., about a century after her alleged death), Mother Shipton was mentioned in print for the first time, in another 17th century chapbook. This is when she was born, at least on paper. Before this first mention, the prophetess hadn't existed except in tradition. However, that in the interim between her claimed birth in the 15th century birth date and 1641, during these silent years, that everything about her passed on by word of mouth, is a possibility. However, my viewpoint is that this would not be expected of a prophet who might have predicted the future of the world so accurately. If she really existed and predicted accurately so many facts, why didn't she receive an immediate recognition? Where are the British records attesting her recognition? Nostradamus, for example, reached wide acknowledgement in the 16th century, as did the astrologer William Lilly in the 17th century. In 1645 the same Lilly published a book including prophecies attributed to Mother Shipton, which are rather similar to those published in 1641. This is additional evidence leading to the opinion that Mother Shipton could be just a legend created in the 17th century, or maybe – if she existed – all the truth regarding her might be contained only in these two 17th century books and nowhere else. This is in light of the fact that the Mother Shipton tale found in the 1684 edition is very different from that in the two earlier books.

The first of Richard Heed editions are found in the British Museum Library. The following shows the text of the 1684 chapbook edition.

The
Life and Death
of
Mother Shipton.

Being not only a true Account of her Strange BIRTH, and most Important Passages of her LIFE, but also of her Prophesies: Now newly Collected. and Historically Experienced, from the time of her Birth, in the Reign of KING HENRY the VII, until this present year 1684, Containing the most Important Passages of State during the reign of these Kings and Queens of England following, *viz.*

Henry the VIII.	King *James*.
Edward the VI.	King *Charles* the I.
Queen *Mary*.	King *Charles* the II.
Queen *Elizabeth*.	Whom God Preserve.

Strangly Preserved amongst other writings belonging to an Old Monastry in *York-shire*, and now published for the Information of Posterity. To which are added some other Prophesies yet unfulfil'd. As also Mr. Folwell's's Predictions concerning the *Turk*, *Pope*, and French King, With Reflections thereupon.

———————

London, Printed for Benj. Harris, at the *Stationers-Armes* and *Anchor* under the *Piazza of the Royal Exchange*. 1684.

Head's book, in the black-letter edition of 1684, sets forth at considerable length that in 1486 a woman named Agatha Shipton lived in a place called "Naseborough" near the Dropping Well in Yorkshire. Her parents died, and she came to poverty. The handsomely guised Devil approached her, made love to her, carried her off on a demon steed, and displayed before her a phantom but apparently real mansion, in which they were married. He promised to give her power over "haile, tempests, with lightning and thunder," the power of traveling from place to place in an instant, and to place the hidden treasures of the earth at her disposal, on certain conditions.

And allured by these promises,

she condescended to all the Devil would have her do. Whereupon he bid her say after him, in this manner: *Raziel ellimiham irammish zirigai Psonthonphanchia Raphael elhaveruna tapinotambecaz mitzphecat jarid cuman hapheah Gabriel Heydonturris dungeonis philonomostarkes sophecord hankim.* After she had repeated these words after him, he plucked her by the Groin, and there immediately grew a kind of Tet, which he instantly sucked, telling her that must be his constant Custom with her, morning and evening; now did he bid her say after him again, *Kametzeatuph Odel Pharaz Tumbagin Gall Flemmegen Victow Denmarkeonto,* having finished his last hellish speech, which the chiefest of his Minions understand not, out of which none but the Devil himself can pick out the meaning; I say, it thundered so horribly, that every clap seemed as if the vaulted roof of Heaven had cracked and was tubling down on her head; and withal, that stately Palace which she thought she had been in, vanished in a trice; so did her sumptuous apparel: and now her eyes being opened, she found herself in a dark dolesome, Wood; a place which from the Creation, had scarce ever enjoyed the benefit of one single Sun-Beam. Whilst she was thinking what course to steer in order to her return, two flaming fiery Dragons appeared before her tyed to a Chariot, and as she was consulting with her self what was best to be done, she insensibly was hoisted into it, and with speed unimaginable conveyed through the Air to her own poor Cottage.

Signs and wonders attended Agatha wherever she went, so that her neighbors were too much afraid of her to persecute her, especially as a winged dragon had once carried her away from the presence of the local magistrate.

In course of time her hellish offspring came into the world. The baby Mother Shipton was

of an indifferent height, but very morose and big bon'd, her head very long, with very great goggling, but sharp and fiery Eyes, her Nose of an incredible and unproportionable length, having in it many crooks and turnings, adorned with many

21

strange Pimples of divers colours, as Red, Blew, and mixt, which like Vapours of Brimstone gave such a lustre to her affrighted spectators in the dead time of the Night, that one of them confessed several times in my hearing, that her nurse needed no other light to assist her in the performance of her Duty: Her cheeks were of a black swarthy Complexion, much like a mixture of black and yellow jaundies; wrinckled, shrivelled, and very hollow; insomuch, that as the Ribs of her Body, so the impression of her Teeth were easily to be discerned through both sides of her Face, answering one side to the other like the notches in a Valley, excepting only two of them which stood quite out of her Mouth, in imitation of the Tushes of a wild Bore, or the Tooth of an Elephant. . . . The Neck so strangely distorted, that her right shoulder was forced to be a supporter to her head, it being propt up by the help of her chin . . . Her Leggs very crooked and mishapen: The Toes of her feet looking towards her left side; so that it was very hard for any person (could she have stood up) to guess which road she intended to stear her course; because she never could look that way she resolved to go.

The lovely creature was put out to nurse at the charge of the parish. Miraculous and unpleasant incidents occurred around her cradle; her attendants were sometimes goaded to exertion by imps in the form of apes. Once Mother Shipton, cradle and all, were missing; sweet harmony from an unknown source was heard; finally the baby and cradle were found three feet up the chimney. As she grew old her foul fiend of a father visited her daily in the form of a cat, dog, bat, or hog. She was sent to school where, says the chronicler,

> her Mistris began to instruct her, as other children, beginning with the Cris-cross-row as they call'd it, showing and naming onely three or four Letters, at first, but to the amazement and astonishment of her Mistris; she exactly pronounced every Letter in the Alphabet without teaching. Hereupon her Mistris, shewed her a Primmer, which she read awel at first sight, as any in the School, and so proceeded in any book was shown her.

Later still Mother Shipton began to tell fortunes, and to foretell the future. High and low flocked to her for information about their private affairs. According to Head, she foretold the visit of Henry VIII to France, the death of Cardinal Wolsey, the downfall of the Catholic power in England, the death of the Duke of Somerset and also that of Lady Jane Grey, and various events in the reigns of Elizabeth, Charles I, Oliver Cromwell, and Charles II. Finally she died, honored and esteemed, and a stone was erected to her memory at Clifton, about a mile from the city of York.

In these very early times Mother Shipton also figured in comedy. An old book in the British Museum Library has the following title page:

The Life of Mother Shipton. A new Comedy. As it was Acted Nineteen days together with great Applause. *Folia Ampla Sybillæ Virg.* Written by T. T.--London, Printed by and for *Peter Lillicrap*, and are to be sold by T. Passinger [Title-page torn here] the real date seems to be about 1660.

The comedy bears a resemblance here and there to Head's narrative. The scene is laid partly in "*Nasebrough Grove in Yorkshire*;" the heroine and prophetess is Agatha Shipton; no daughter Ursula appears in it at all. On page 15 a village crier is made to announce "O Yes, if any man or woman, in City, Town or Country can tell me tydings of *Agatha Shipton*, the daughter of *Solomon Shipton* Ditch digger lately deceased, let them bring word to the Cryer of the village, and they shall be well rewarded for their pains."

Agatha marries the devil, as in other versions of the story, but cheats him at the last:

[Soft Musick and an Angel descends with a Book.]
Shipton despair not but in hope grow strong,
Thou shalt find Mercy though thou hast done wrong;

24

Road ore this book and in it thou shalt find

The summe of thy aspire to free thy mind

From fear, thy soul secure from harm

Of any Devils! 'tis a happy charme!

Pluto enters with "all the Devils," and finding Agatha Shipton

released from their power exclaims:

Was ever Devil gull'd so

  "VVell lets descend and all Hell shall howl

  This full fortnight for losse of Shipton's soul.

    "[Exeunt with horrid Musick].

"Shipton. So let them roare.

  VVhilst I do all their Hellish Acts despise

  The higher powers make me truly wise.

18th century woodcut,
from *The History of Mother Shipton*

CHAPTER 3

THE END OF THE WORLD[2]

There are different histories about Mother Shipton, while some people claim she never existed and even her existence is a forgery. Perhaps one of the most famous prophecies attributed to her is the following predicting catastrophes for 1881. This is the year in which, according to this celebrated Yorkshire prophetess, the world would come to an end:

> The world to an end shall come,
> In eighteen hundred and eighty one.

This prophecy was published for the first time 19 years before 1881. At that time, even the "scientific and skeptical mind" (as an old book says) showed curiosity. In the 20th century, this

---

[2] This chapter was based on the 1881 William H. Harrison Book, Mother Shipton: *The Yorkshire Sibyl Investigated.*

prophecy was changed and had its final year changed to 1991. This, as well as other prophecies, was said to have been copied from records of unimpeachable antiquity in the British Museum Library; these prophecies in some cases were reproduced in alleged *fac-simile*.

According to Charles Hindley, the man who pointed out the end of the world in the 1881 prophecy, these words were published initially in 1448 and then reprinted in 1641. Not true.

The end of the world had already been predicted many times. The alchemist and astrologer Arnaldo di Villanova, for example, pointed to the end as in 1335. The astrologer Giovanni Carion announced the apocalypse for 1532, and in Germany the astrologer Carl Stoffer announced it for 1524.

Prophecies about the end of the world have always had some influence. In *Supputatio annorum mundi*, Luther presents a chronology of the world pointing out its end as in the 16th century. According to him the world's timespan would be six thousand years and in "this year (1540) the number of years is exactly 5500, but because the three-day period of dead Christ was two days and a half, the end should arrive in 1550..."

The 2012 end is found in many books of prophecies printed today, but this is not new. The first to call attention to the 2012 end of the world seems to be a certain man living centuries ago called Aretius who, based on the *Adventus Domini*, calculated the end for 2012.

Whiston predicted that the world would be destroyed on the 13th October, 1736. Then, crowds of people left London to see, from neighboring fields, the destruction of the city, which was to be "the beginning of the end."

Many people today still believe that European fanatics predicted the end of the world for 999 or 1000. The truth is that this was a myth created a few centuries after the year 1000, as the French Historian Le Goff explains in one of his books about Middle Ages. Actually, this end-of-the-world fear did not exist then, mainly because the common man or peasant who lived in the country at that time did not even know the year in which he lived or the year he was born. The text below was one of the lies that contributed to the 999-1000 end-of-world myth:

The scene of the last judgment was expected to be at Jerusalem. In the year 999, the number of pilgrims proceeding eastward, to await the coming of the Lord in that city, was so great that they were compared to a desolating army. Most of them sold their goods and possessions before they quitted Europe, and lived upon the proceeds in the Holy Land. Buildings of every sort were suffered to fall into ruins. It was thought useless to repair them when the end of the world was so near. Many noble edifices were deliberately pulled down. Even churches, usually so well maintained, shared the general neglect. Knights, citizens, and serfs, travelled eastwards in company, taking with them their wives and children, singing psalms as they went, and looking with fearful eyes upon the sky, which they expected each minute to open, and to let the Son of God descend in glory.[3]

A panic occurred in Leeds in 1806, during which many in their fear "got religion" for a time, and indulged in a temporary repentance. In a village close by, a Yorkshire hen had been laying eggs inscribed, "Christ is coming." Eventually the writing was discovered to be in corrosive ink, and the trick found out by which observers were made to believe that the hen laid them in that condition.

The *Pall Mall Gazette* of April 14th, 1879, says that the Mid-Somerset people believed Mother Shipton to have prophesied that on Good Friday, 1879, Ham Hill, near Yeovil, would be swallowed up at 12 o'clock by an earthquake, and Yeovil itself

---

[3] Mackay's *Popular Delusions*. London: 1869. Vol. i, p. 222.

visited by a tremendous flood. Some people actually left the locality with their families to avoid the calamity; others made various preparations for it. On that Good Friday large numbers of people flocked to the vicinity of Ham Hill, expecting to see it swallowed up, but they were disappointed.

The following is the most largely circulated form of one of Shipton's reputed prophecies, which of late years has been in the public mind. It's quoted from page 450 of *Notes and Queries*, December 7th, 1872.

ANCIENT PREDICTION

(Entitled by Popular tradition 'Mother Shipton's Prophecy,')
Published in 1448, republished in 1641.[4]

Carriages without horses shall go,
And accidents fill the world with woe.
Around the world thoughts shall fly
In the twinkling of an eye.
The world upside down shall be
And gold be found at the root of a tree.
Through hills man shall ride,
And no horse be at his side.
Under water men shall walk,
Shall ride, shall sleep, shall talk.
In the air men shall be seen,
In white, in black, in green;
Iron in the water shall float,
As easily as a wooden boat.
Gold shall be found and shown
In a land that's now not known.

---

[4] This prophecy was never published these years; it was created a few years before 1881.

Fire and water shall wonders do,
England shall at last admit a foe.
The world to an end shall come,
In eighteen hundred and eighty one."

This prophecy has been printed many times in 20th century books. Some of these profess to give her authentic history with the marvelous elements sifted out; others include miraculous incidents.

The following account of her life, as adapted to the more sober-minded readers of the present century, is summarized by William Harrison from a book entitled *Mother Shipton and Nixon's Prophecies*, compiled from original and scarce editions by S. Baker, published in 1797, by Denley, Gate Street, Lincoln's Inn Fields, London. The pamphlet gives information about the life of Nixon, a Cheshire prophet, also about Ursula Shipton, for Ursula is the real name of our heroine. She is stated by Baker to have been born in July 1488, in the reign of Henry VII, near Knaresborough, Yorkshire. She was baptized by the Abbot of Beverley, by the name of Ursula Sonthiel. "Her stature," adds her biographer, "was larger than common, her body crooked, her face frightful; but her understanding extraordinary."

Baker states that she was a pious person, who at the age of twenty-four was courted by one Toby Shipton, a builder, of

Skipton, a village four miles north of York; soon afterwards they were married. She became known as Mother Shipton, and acquired fame by means of her extraordinary predictions.

Some other prophecies are attributed to her. For example, when Cardinal Wolsey intended to remove his residence to York, she announced that he would never reach that city. The Cardinal sent three lords of his retinue in disguise, to inquire whether she had made such a prediction, and to threaten her if she persisted in it. She was then living in a village called Dring Houses, a mile to the west of the city. The retainers, led by a guide named Beasly, knocked at the door.

"Come in, Mr. Beasly, and three noble lords with you," said Mother Shipton.

She then treated them civilly, by setting out cakes and ale before them.

"You gave out," said they, "the Cardinal should never see York."

"No," she replied, "I said he might see it, but never come to it."

They responded, "When he does come, he'll surely burn thee."

"If this burn," said the Reverend Mother, "so shall I."

She then cast her linen handkerchief into the fire, allowed it to remain in the flames a quarter of an hour, and took it out unsinged.

One of her awe-stricken visitors then asked what she thought of him.

She answered, "The time will come, my lord, when you shall be as low as I am, and that is low indeed."

This was judged to be verified when Thomas Lord Cromwell was beheaded.

Cardinal Wolsey, on his arrival at Cawood, ascended the Castle Tower, and while viewing York, eight miles off, vowed he would burn the witch when he reached there. But ere he descended the stairs, a message from the King demanded his presence forthwith, and while on his journey to London, he was taken ill and died at Leicester.

She foretold the destruction by tempest of the Ouse Bridge and Trinity Church, York, in the following mystical language: "Before Ouze Bridge and Trinity Church meet, what is built in the day shall fall in the night, till the highest stone of the church be the lowest stone of the bridge."

Baker's booklet passed through two editions in 1797. He alleges that some of her prophecies therein were copied from an "original scroll delivered by her to the Abbot of Beverley; privately preserved in a noble family for many years, and lately discovered among other curious and valuable manuscripts." He states that she foretold the time of her death, and that after taking solemn leave of her friends she departed, with much serenity, A.D. 1651, when upwards of seventy years of age. A stone monument was erected to her memory on the high North Road, between the villages of Clifton and Skipton, about a mile from York. The monument represents a woman upon her knees, with her hands closed before her, in a praying posture, and "stands to be seen there to this day," (1797). The following is said to have been her epitaph:

Here ly's she who never ly'd
Whose skill often has been try'd
Her Prophecies shall still survive,
And she keep her name alive.[5]

Mother Shipton 1648 century woodcut

[5] Even though it seems this epitaph may exist in Clifton, there is however no proof that the stone in question marks her grave or that she was buried there.

CHAPTER 4

*M*any old books were printed in the 17th and 18th century about Mother Shipton, but as one can see, these books did not include any prophecy regarding our time. They only had predictions referring to past events, such as the Spanish Armada, the Great Fire of London, etc. More recently, in the 19th century, new forgeries attributed to her were printed, with the same happening in the 20th century. It's interesting to look at these first publications as examples of the kind of prophecy attributed to Mother Shipton at each time.

The earliest known record presently in existence relating to Mother Shipton is a pamphlet in good preservation, dated 1641, presented to the British Museum by King George III. The following is a reprint of the whole of it:

THE
PROPHESIE
OF
MOTHER *SHIPTON*
In the Raigne of King
Henry the Eighth.

Fortelling the death of Cardinall *Wolsey*, the Lord *Percy*
and others, as also what should happen in
insuing times.

LONDON,
Printed for *Richard Lownds*, at his Shop
adjoyning to Ludgate. 1641.

The Prophecy of Mother *Shipton;*
in the Reign of King *Henry*
the Eighth.

HEN she heard King *Henry* the eighth should be King, and Cardinal *Wolsey* should be at *York*, she said that Cardinal *Wolsey* should never come to *York* with the King, and the Cardinal hearing, being angry, sent the Duke of *Suffolk*, the Lord *Percy*, and the Lord *Darcy* to her, who came with their men disguised to the Kings house near *York*, where leaving their men, they went to Master *Besley* to *York*, and desired him to go with them to Mother *Shiptons* house, where when they came they knocked at the door, she said Come in Master *Besley*, and those honourable Lords with you, and Master *Besley*, would have put in the Lords before him, but she said, come in Master *Besley*, you know the way, but they do not. This they thought strange that she should know them, and never saw them ; then they went into the house, where there was a great fire, and she bade them welcome, calling them all by their names, and sent for some Cakes and Ale, and they drunk and were very merry. Mother *Shipton*, said the Duke, if you knew what we come

39

The Prophecy of Mother *Shipton;*
in the Reign of King *Henry* the Eighth.

HEN she heard King *Henry* the eighth should be King, and Cardinal *Wolsey* should be at *York*, she said that Cardinal *Wolsey* should never come to *York* with the King, and the Cardinal hearing, being angry, sent the Duke of *Suffolk*, the Lord *Percy*, and the Lord *Darcy* to her, who came with their men disguised to the Kings house near *York*, where leaving their men, they went to Master *Besley* to *York*, and desired him to go with them to Mother *Shiptons* house, where when they came they knocked at the door, she said Come in Master *Besley*, and those honourable Lords with you, and Master *Besley*, would have put in the Lords before him, but she said, come in Master *Besley*, you know the way, but they do not. This they thought strange that she should know them, and never saw them ; then they went into the house, where there was a great fire, and she bade them welcome, calling them all by their names, and sent for some Cakes and Ale, and they drunk and were very merry. Mother *Shipton*, said the Duke, if you knew what we come about, you would not make us so welcome, and she said the messenger should not be hang'd ; Mother *Shipton*, said the Duke, you said the Cardinal should never see *York* ; Yea, said she, I said he might see *York*, but never come at it ; But said the Duke, when he comes to *York* thou shalt be burned ; We shall see that, said she, and plucking her Handkerchief off her head she threw it into the fire, and it would not burn, then she took her staff and turned it into the fire, and it would not burn, then she took it and put it on again; Now (said the Duke) what mean you by this? If this had burned (said she) I might have burned. Mother *Shipton* (quoth the Duke) what think you of me ? my love said she, the time will come you will be as low as I am, and that's a low one indeed. My Lord *Percy* said, what say you of me ? My Lord (said she) shoe your Horse in the quick, and you shall do well, but your body will be buried in *York* pavement, and your head shall be stolen from the bar and carried into *France*. Then said the Lord *Darcy*, and what think you of me ? She said, you have made a great Gun, shoot it off, for it will do you no good, you are going to war, you will pain many a man, but you will kill none, so they went away.

Not long after the Cardinal came to *Cawwood*, and going to the top of the Tower, he asked where

York was, and how far it was thither, and said that one had said he should never see *York* ; Nay, said one, she said you might see *York*, but never come at it. He vowed to burn her when he came to *York*. Then they shewed him *York*, and told him it was but eight miles thence; he said that he will be soon there: but being sent for by the King, he died in the way to *London* at *Leicester* of a lask ;[1] and *Shiptons* wife said to Master *Besley*, yonder is a fine stall built for the Cardinal in the Minster, of Gold, Pearl, and precious stone, go and present one of the pillars to King *Henry*, and he did so.

Master *Besley* seeing these things fall out as she had foretold, desired her to tell him some more of her Prophesies; Master, said she, before that *Owes*[2] Bridge and Trinity Church meet, they shall build on the day, and it shall fall in the night, until they get the highest stone of Trinity Church, to be the lowest stone of *Owes* bridge, then the day will come when the North shall rue it wondrous sore, but the South shall rue it for evermore ; When Hares kindle on cold hearth stones, and Lads shall marry ladies, and bring them home, then shall you have a year of pining hunger, and then a dearth without Corn ; A woeful day shall be seen in *England*, a King and Queen, the first coming of the King of

[1] LASK.—A laxity, a looseness or flux,
[2] OWES.—*i.e.,* Ouse,

42

Scots shall be at *Holgate* Town, but he shall not come through the bar, and when the King of the North shall be at *London* Bridge, his Tail shall be at *Edenborough* ; After this shall water come over *Owes* bridge, and a Windmill shall be set on a Tower and an Elm-tree shall lay at every mans door, at that time women shall wear great hats and great bands, and when there is a Lord Mayor at *York* let him beware of a stab ; When two Knights shall fall out in the Castle yard, they shall never be kindly all their lives after ; When all *Colton* Hagge hath born seven years Crops of corn, seven years after you heard news, there shall two Judges go in and out at *Mungate* bar.

> *Then Wars shall begin in the spring,*
> *Much woe to* England *it shall bring :*
> *Then shall the Ladies cry well-away,*
> *That ever we liv'd to see this day,*

Then best for them that have the least, and worst for them that have the most, you shall not know of the War over night, yet you shall have it in the morning, and when it comes it shall last three years, between *Cadron* and *Aire* shall be great warfare, when all the world is as a lost, it shall be called Christs cross, when the battle begins, it shall be where Crookbackt *Richard* made his fray, they shall say, To warfare for your King for half a crown a

43

day, but stir not (she will say) to warfare for your King, on pain on hanging, but stir not, for he that goes to complain, shall not come back again. The time will come when *England* shall tremble and quake for fear of a dead man that shall be heard to speak, then will the Dragon give the Bull a great snap, and when the one is down they will go to *London* Town; then there will be a great battle between *England* and *Scotland*, and they will be pacified for a time, and when they come to *Brammammore*, they fight and are again pacified for a time, then there will be a great Battle at *Knavesmore*, and they will be pacified for a while; Then there will be a great battle between *England* and *Scotland* at *Stoknmore*; Then will Ravens sit on the Cross and drink as much blood of the Nobles, as of the Commons, then woe is me, for *London* shall be destroyed for ever after; Then there will come a woman with one eye, and she shall tread in many mens blood to the knee, and a man leaning on a staff by her, and she shall say to him, What art thou; and he shall say, I am King of the *Scots*, and she shall say, Go with me to my house, for there are three Knights, and he will go with her, and stay there three days and three nights, then will *England* be lost; and they will cry twice of a day *England* is lost; Then there will be three Knights in *Petergate* in *York*, and the one shall not know of the

other ; There shall be a child born in *Pomfret* with
three thumbs, and those three Knights will give him
three horses to hold, while they win *England*, and
all Noble blood shall be gone but one, and they
shall carry him to Sheriff *Nuttons* Castle six miles
from *York*, and he shall die there, and they shall
choose there an Earl in the field, and hanging their
horses on a thorn, and rue the time that ever they
were born, to see so much bloodshed ; Then they
will come to *York* to besiege it, and they shall keep
them out three days and three nights, and a penny
loaf shall be within the bar at half a crown, and
without the bar at a penny ; and they will swear if
they will not yield, to blow up the Town walls.
Then they will let them in, and they will hang up
the Mayor, Sheriffs and Aldermen, and they will go
into Crouch Church, there will three Knights go in,
and but one come out again, and he will cause
Proclamation to be made, that any man may take
House, Tower, or Bower for twenty one years,
and whilst the world endureth, there shall never be
warfare again, nor any more Kings or Queens, but
the Kingdom shall be governed by three Lords, and
then *York* shall be *London ;* and after this shall be
a white Harvest of corn gotten in by women.
Then shall be in the North, that one woman shall
say unto another, mother I have seen a man to-day,
and for one man there shall be a thousand women,

there shall be a man sitting upon St. *James* Church hill weeping his fill; and after that a ship come sailing up the Thames till it come against *London*, and the Master of the ship shall weep, and the Mariners shall ask him why he weepeth, being he hath made so good a voyage, and he shall shall say; Ah what a goodly City this was, none in the world comparable to it, and now there is scarce left any house that can let us have drink for our money.

> *Unhappy he that lives to see these days,*
> *But happy are the dead* Shiptons *wife says.*

FINIS.

CHAPTER 5

William Lilly, the famous 17th century British astrologer, published *A Collection of Ancient and Moderne Prophesies.* . . . . *London, Printed for John Partridge and Henry Blunden, and are to be solde at the Signe of the Cock, in Ludgate Streete, and the Castle in Cornehill, 1645.* This book is the second oldest source of information after the 1641 book and it contains Shiptons prophecy, *"after the most exact copy."* Lilly's version may (or may not be) the more trustworthy of the two, due to the care professedly exercised by Lilly in the selection. Lilly's version is found in the British Museum Library.

The 1645 Lilly edition and the 1641 were compared and they agree tolerably closely. Here and there Lilly's version contains trifling additions not in the earlier pamphlet. For instance, it

---

[6] This chapter was based on the 1881 William H. Harrison Book, Mother Shipton: *The Yorkshire Sibyl Investigated*

says that after Mother Shipton told Lord Percy that his body would be buried in York pavement and his head carried into France, "They all laughed saying, that would be a great lop between the Head and the Body."

But this 1645 pamphlet is of exceeding interest, because it shows that nearly all the alleged prophecies of Mother Shipton published in these earlier records had been fulfilled before 1645, that is to say, they have been fulfilled more than 200 years ago; Lilly's reprint sets forth the following points in relation to the fulfillment of the prophecies printed in the last chapter:

I. That the Duke of Suffolk had been beheaded.
II. That Lord Percy had risen in rebellion in the North, that he had been beheaded and that his body was buried in York; also that his Head was stoln away and carried into France. Temp. Eliz. R.
III. That Trinity steeple in York had been blown down in a tempest, and Ouse bridge broken down by a great flood; also that the repairs made in the day fell down in the night, till they, remembering the prophecy, made the highest stone of the steeple the foundation of the bridge, and then the work stood. By this was partly verified another of Mother Shipton's sayings, "that her maid should live to drive her cow over Trinity steeple."
VI. The prophecy about the North rueing it "wondrous sore," is supposed to refer to the suppression of religious houses, and "at the Lord William Howard's house at Naworth, a Hare came and kinnell'd in his Kitchin, upon the hearth."
V. As to the King of Scots at Holgate Town. When King James arrived at Holgate, such a multitude had assembled that he was forced to ride another way. His children were in Edinburgh.
VI. As to the prophecy about the water over Ouso bridge and the windmill on a tower, water was carried into York through "boared Elmes," and a windmill drew up the water at Conduit House.
VII. A Lord Major whose house was in the Minster yard in York, was killed with three stabs.

VIII. "Sir *T. Wentworth* and Sir *John Savil*, in choosing Knights for the shire, in the Castle-yard at Yorke, did so fall out, that they were never well reconciled."

IX. "Colton hag in her time was a Woodland ground full of trees, which bore Corn seven years, and the seventh yeer after that was the yeer of the coming in of the *Scots*, and their taking of *Newcastle*."

X. "In the yeer 1616 the two Judges of Assize went out at a gate in York, where never any Judges were known to go out before or since."

XI. About wars beginning in the spring, King Charles raised an army in the spring of 1639, after which many ladies lost their husbands, and people were so taxed it was worst for those who had the most.

XII. "Calder and Are" are two Yorkshire rivers, and "Are passeth through Craven."

XIII. Where "Crookback Richard made his fray." This, says the chronicler, refers to "Neer Leicester, where Richard the third was slain in battel, there Colonel Hastings was one of the first in arms, endeavouring to settle the Commission of Array, in opposition to others, that were then setling the Militia."

XIV. "1642. Two shillings and sixpence was publikely promised by many Lords for the King's use, to pay one Horseman a day's wages."

XV. Many Welsh and Irish were killed in the war.

XVI. The prophecy about quaking for fear of dead man, not fulfilled.

XVII. War between England and Scotland not fulfilled. "Brammish is a river in Northumberland."

XVIII. A child had been "credibly reported" to have been born at Pomfret with three thumbs.

XIX. The prophecy of the siege of York and its accompanying incidents not fulfilled.

XX. The prophecy about London not fulfilled.

The foregoing catalogues nearly all Mother Shipton's prophecies as having been dated to be fulfilled before 1645. That of the mariner in the Thames weeping for malt liquor in the partly destroyed city may more particularly be supposed to yet remain for fulfillment, but Mr. Baker, the writer of her 1797 biography, claims that this last one describes the results of the Great Fire of London in 1666, which left not one house between the Tower and the Temple. This fire, at all events, occurred long after

49

Mother Shipton's death and the publication of her alleged prophecy.

The third copy in point of antiquity, of Mother Shipton's Prophecies in the British Museum, is a black-letter pamphlet, published in 1663, "Printed by *T. P.* for *Fr. Coles*, and are to be sold at his shop at the signe of the Lambe in the Old-Baily, neare the Sessions House, 1663." It is entitled *Mother Shipton's Prophesies: with Three and XX more, all most terrible and wonderfull, Predicting strange alterations to befall this Climate of England.*

This version agrees closely with Lilly's, but the latter is rather more complete, and is in a better state of preservation. The 1663 edition, however, ends with the following couplet, not given by Lilly:

> In the world old age this woman did fore-tell,
> Strange things shal hap, which in our time have fell.

Mother Shipton's prophecies, therefore, were generally recognized as having been fulfilled before the middle of the 17th century.

CHAPTER 6

I n the previous chapters, different old books were introduced and brought together to give some background for a critical examination.

The 17th century records in the British Museum Library, in relation to Mother Shipton, agree closely with each other, but none of them contain the lines ending with the too-often celebrated couplet:

> The world to an end shall come,
> In eighteen hundred and eighty one.

These lines and the notorious prophecy about the end of the world in 1881 were fabricated in the 19th century by Mr. Charles Hindley. The editor of *Notes and Queries* says, in the issue of that journal dated April 26th, 1873:

---

[7] This chapter was based on the 1881 William H. Harrison Book, Mother Shipton: *The Yorkshire Sibyl Investigated*

Mr. Charles Hindley, of Brighton, in a letter to us, has made a clean breast of having fabricated the Prophecy quoted at page 450 of our last volume, with some ten others included in his reprint of a chap-book version, published in 1862.

Most of the precise details about the birth, life, and death of Mother Shipton are fabrications which have been reproduced time after time in chapbooks. There is no absolute evidence that any one of the details is true, but there may be some foundation for the incident narrated about Cardinal Wolsey.

The details presented by the Richard Head book, which have interested the public for 200 years, are fabrications. "Richard Head, *gentleman*," drew the contents of every page of his book from his own inner consciousness. His preface to the oldest edition of his work extant (16 4) is amusing, and among other items sets forth as follows, how he obtained and dealt with the alleged Shipton manuscript:

> Many old Manuscripts and rusty Records I turned over, but all in vain; at *last* I was Informed by a Gentleman (whose Ancestors by the Gift of King *Henry* the Eighth, enjoyed a. *Monastary* in these parts) that he had in his keeping some Ancient Writings which would in that point satisfie my desire, were they not so Injured by Time, as now not legible to Read; however, I not despairing to find out their meaning, with much Importunity desired to have a sight of them; which having

obtained, I took of the best Galls I could get, beat them grosly, and laid them to steep one day in good White-wine, that done, I distilled them with the Wine; and with the Distilled Water that came off them, I wetted handsomely the old Letters, whereby they seemed as fresh and fair as if they had been but newly Written.

From the above, it would appear that even in Head's day there was a desire for earlier manuscripts about Mother Shipton.

Richard Head, who has so long misdirected the thoughts of large numbers of people, was the son of a minister in Ireland. Head's father was massacred "with many thousands more" in 1641. Mrs. Head then brought her son to England, and he completed his studies at Oxford. He could not afford to remain until he obtained a degree, so turned bookseller. He married, and soon afterwards became a ruined man, in consequence, says Erskine Baker, "of two pernicious passions, viz., poetry and gaming, the one of which is for the most part unprofitable, and the other almost always destructive." He retired to Ireland, where, in 1663, he wrote his only dramatic piece, *Hic et Ubique*, by which piece he acquired great reputation, and some money. As a literary man he had several ups and down in the world; his writings had a strong tinge of indecency. He was drowned in the year 1678, while crossing to the Isle of Wight.

The other piece of fiction of high antiquity, relating to our heroine, is the comedy of the *Life of Mother Shipton*, which is said to have been acted nine days with great applause. The author was one T. Thompson. The British Museum authorities consider the date of the Mother Shipton comedy to be about 1660, so it ranks with the earliest existing narratives relating to the subject. In the *Lives and Characters of the English Dramatick Poets . . . . . First begun by Mr. Langbain, improv'd and continued down to this Time, by a Careful Hand, London: Printed for William Turner, at the White Horse, without Temple Bar, 1699*, Langbain describes Thomas Thompson as

> A Poor Plagiary, that could not disguise or improve his Thefts. Those two following Plays go under his Name; viz.
> The *English Rogue*, a Comedy, 4*to*. 1688, acted (says the Title) before several Persons of Honour, with great Applause, and dedicated to Mrs. *Alice Barrett*.
> *Mother Shipton, her Life*; 4*to*. The Author hereof says, 'twas acted Nine Days together, with great Applause. Plot from a Book so called in the Prose, 4*to*., but most of the Characters and Language from *The City Madam*, and *The Chast Maid of Cheapside*.

Thompson's play of The English Rogue, was also dramatized from a book by Richard Head, for whose dubious writings Thompson, therefore, seems to have had admiration.

There may be other ancient versions of Mother Shipton's prophecies, but it seems that there isn't any other known older than that of 1641. *Notes and Queries*, of July 25th, 1868, contains a letter from an anonymous writer, making mention of some old editions which may be in other collections. His exceptionally valuable remarks about Mother Shipton and her history are abridged:

Although the fact of the existence of Mother Shipton rests wholly upon Yorkshire tradition, she can scarcely be regarded as a myth. According to the tradition, the place of her birth was on the picturesque banks of the river Nidd, opposite to the frowning towers of Knaresborough Castle, and at a short distance from St. Robert's Cave--a spot famous for mediæval legends and modern horrors. She first saw the light a few years after the accession of Henry VII. It was not until fourscore years after her death that any account of her extraordinary predictions was recorded in print. A few years before the breaking out of the Civil War, King Charles I frequently passed through Yorkshire, and perhaps the prophecies of the Yorkshire witch then prevalent in the county, captivated the imagination of some follower of the Court, who on his return to London concocted the first pamphlet. It soon became popular, and the following year two reprints appeared, with some additional prophecies. In 1643 a third edition was published, which was followed by two others a few years afterwards. In 1662 and 1663, after the Restoration, the tracts already described were reprinted with some additional matter, and in 1667 the notorious Richard Head, author of several works of a loose description, invented her biography, and gave to the world a new version of her prophecies. This production has been accepted by the popular taste as the authentic history of the Yorkshire witch, and has been reprinted and sold in all parts of the kingdom. Drake, the historian of York, states that Cardinal Wolsey never came nearer to York than Cawood, which makes good a prophecy of Mother Shipton. "I should not have noticed this idle story," he adds, "but that it is fresh in the mouths of our country people at this day; but whether it was a real prediction, or raised after the event, I shall not take upon me to determine. It is more than probable, like all the rest of these kind of tales, the accident gave occasion to the story." (See Eboracum, p. 450, and get date of it). In a *History of Knaresborough*, published by Harcourt about a hundred years ago, Mother Shipton's traditionary prophecies are described as being still familiar in her native town. The much mutilated sculptured stone near Clifton, Yorkshire, universally called "Mother Shipton," was the figure of a warrior in armour, which had been a recumbent monumental statue; it was probably brought from the neighbouring Abbey of St. Mary, and placed upright as a boundary stone. It has been removed to the museum of the Yorkshire Philosophical Society.

After taking away palpable fiction, we are left face to face with three of the earliest editions of Mother Shipton's prophecies. These agree closely with each other in their details, the variations being few and unimportant. They appear to have been written seriously and with a desire for truth, in which they differ from the Shipton literature of the last 200 years.

A critical examination of the oldest record, reprinted here, reveals that the first part was written by one man, and the second part by another; the former was the most able of the two.

56

The latter part consists of Besley's statements, evidently made originally in doggerel verse, but set by the printer, for the most part, in prose. The rhymes can be traced.

Lilly's 1645 version is the best of the three, and it preserves more of Master Besley's rhymes in their original form. For instance, the 1641 edition contains the following lines:

> Then Warres shall begin in the spring,
> Much woe to England it shall bring:
> Then shall the Ladyes cry well-away,
> That ever we liv'd to see this day.

Lilly's edition gives the following more complete quotation from an older version:

> The North shall rue it wondrous sore,
> But the South shall rue it for evermore.
> When wars shall begin in the spring
> Much wo to England it wild bring:
> Then shall the Ladies cry well a-day,
> That we ever liv'd to see this day.
> Then best for them that have the least
> And worst for them that have the most.

Not only is there this internal evidence of the pamphlets being more or less true copies of earlier records, but Lilly, in his

Collection of Ancient and Modern Prophecies, published in 1645, makes this direct statement in the "introduction to the reader:"

> Mother Shipton's" [prophecy] "was never yet questioned either for the verity or antiquity; the North of England hath many more of hers.

Has such a person as Mother Shipton ever lived?

Cardinal Wolsey was at Cawood in 1530, and the earliest record in existence of Mother Shipton is dated 1641, leaving a gap of 111 years between the chief incident of her career and the oldest record thereof. But Lilly in 1645 speaks of various earlier records of her prophecy being then in existence, and of the facts being in his day undisputed. Some of those older records, which between 1641 and 1663 were reprinted with much fidelity, might possibly have been issued, if not in the lifetime of the sibyl herself, at all events in the lifetime of some of those who dwelt in York when the occurrences took place. After Cardinal Wolsey's death, Mother Shipton told Master Besley to take a jeweled pillar out of York Cathedral and to present it to Henry VIII. It might be asked how Master Besley could do this at the mere instigation of an old woman, and without the consent of the Archbishop. But history shows that the See of York was vacant for nearly a year after Cardinal Wolsey's death, so that while it was in the charge of underlings, at a time when Henry VIII began to seize church property in all directions. Thus this

Mr. Besley may have had the power to do what is recorded of him. Besley's name is spelt "Beasley"'" in Lilly's reprint of the Shipton prophecy, and I find in Drake's *Eboracum* that in the year 1486 a John Beasley was one of the Sheriffs of York.

The admirer of Mother Shipton may have been his son; at all events people of that name *were* living in York before the incident with Cardinal Wolsey is said to have occurred.

In 1539, Richard Layton, Dean of York, pawned some of the jewels of the Cathedral, which is a corroborative illustration of the treatment of church property at that period.

Not so very long after the event, then, a clear record of the interview of Mother Shipton with the three lords found its way into print, and the writer lengthened the narrative by tacking some of Master Besley's doggerel verses to the end of it. If there were no truth in the story, it was one which would have given much offence to the immediate descendants of the noblemen whose names had been so freely used in public.

Lilly, as already stated, raises no question that Mother Shipton existed, and says that in his time the authenticity of her prophecies was undisputed.

17th century image from a chapbook about Mother Shipton as a witch

CHAPTER 7

There are different mysteries regarding Mother Shipton and her prophecies. According to some writers, one is related to the following prophecy, which was attributed to her and is found in a tombstone in England.

When pictures look alive with movements free,

when ships, like fish, swim beneath the sea,

when men, outstripping birds shall scan the sky,

then half the world deep drenched in blood shall lie.

While some books say Mother Shipton made the above prediction and others explain its similarity to other prophecies she announced, they thus assert the above prophecy is probably hers too. But, as we will see, the prophetess never predicted such a thing.

A tombstone is said to exist in a certain Kirby Cemetery in Essex in England, with this prediction inscribed on it. Before locating the tombstone, the issue begins with the cemetery. Does it exist? If so where is it? The only Kirby in Essex is a charming town called Kirby-le-Soyen. There is also a place called Kirby Cross, which is a modern development of part of the old town. Based on this information, if a book mentions a prophecy related to Kirby Cemetery, it refers to Kirby-le-Soyen, not to Kirby Cross. Even though the aforementioned cemetery may exist, it seems that nobody has ever found the inscription mentioned in some books. Most probably many writers just copied the prophecy from other books without caring really to visit the cemetery to verify whether the tombstone exists.

After looking for some information, a person living in Kirby-le-Soken, who had done some research about this matter, gave me an interesting explanation, which will help to understand the cemetery prophecy issue:

> It is indeed said to be an inscription from somewhere within the church or churchyard at Kirby-le-Soken. The problem is that it's never been located, despite much interest for many years. I have heard that the inscription was once on a memorial in the church itself and that it is on a gravestone in the churchyard - neither has ever been located.

My mother helped catalogue the gravestones in the churchyard in the 1990s but nothing was ever found.

What I can say is that most of the very old gravestones are now so badly eroded that they cannot now be read so it is just possible that the inscription, if it ever existed, is now lost forever.

But let's not jump into conclusions quickly, finishing here by saying the famous tombstone prophecy is fake. One could also try to discover the first time this wonderful prophecy was printed. If I could find this prophecy printed in a 16th century book, for example, that would be enough to convince me. At least, even if the tombstone did not exist, we would have an old book attesting the prophecy existed. However, after checking many 16th and 17th century books, I never found that prophecy, which makes me think this one is probably a 20th century forgery invented in the beginning of the century, before or around the First World War.

A similar prediction about the supposed cemetery prophecy might perhaps exist in a 15th century book called *Liber Elemonsinatoris*. However, I haven't been able to find such a book. If this book exists (probably as a manuscript in Latin), and this prediction for submarines and airplanes is really found there, it is probably the source of the copied prophecy. I would risk this vaticination doesn't exist, though. I'd bet this is another

20th century creation. Concerning the Kirby-le-Soyen tombstone and its prediction, it might even have existed; on the other hand, unfortunately it seems not possible to be sure about it anymore.

CHAPTER 8

There is also another prophecy attributed to Mother Shipton, which was initially published in 1995 in an Australian magazine. The article says this prophecy had been found in Australia, in an old manuscript scroll hidden in an library. This prophecy might be a forgery too, because so far I could not find evidence about its existence. Some Australians assert this document is owned by the New South Wales library, in Australia. Others claim it is located in the Australian Historical Archives, yet others say the prophecy was in scrolls found in jars in Mitchell Library, in Sidney. According to some, the Shiptons left Yorkshire and went to Australia in the 1880s, and established themselves around a tin mine near Cooktown, named Shipton's Flat, but this again is not definitive.

Some books and the Internet might be creating further confusion regarding this prophecy, because they published it as an old and real prophecy, without also revealing to the reader it was printed for the first time ever in 1995, in Australia. So, could it be just a well-elaborated hoax?

Below you will find the prophecy as it was published in the Australian magazine called *Nexus* (Volume 2, 24, February-March 1995), in 1995. Part of this prophecy (those words in bold and in italic) is really older and can be found in books printed in the 19th and 20th century, with a few variations.

The first lines in the beginning of the prophecy, which are in bold and italic, are those that had been previously attributed to Mother Shipton in the 19th century. But as you will see, after a certain point in the text, they slide into what seems to be a 20th century fakery. I'd risk and say this is a typical forgery. This kind of forgery, mixing older words with new ones, was done a few times in Europe in the past. Naturally, I am not able to state who composed it, and I obviously don't believe the magazine editor published it thinking the same I do; he probably believed it was a real prophecy.

AND NOW A WORD, IN UNCOUTH RHYME

OF WHAT SHALL BE IN FUTURE TIME

THEN UPSIDE DOWN THE WORLD SHALL BE

66

AND GOLD FOUND AT THE ROOT OF TREE

ALL ENGLAND'S SONS THAT PLOUGH THE LAND

SHALL OFT BE SEEN WITH BOOK IN HAND

THE POOR SHALL NOW GREAT WISDOM KNOW

GREAT HOUSES STAND IN FARFLUNG VALE

ALL COVERED O'ER WITH SNOW AND HAIL

A CARRIAGE WITHOUT HORSE WILL GO

DISASTER FILL THE WORLD WITH WOE.

IN LONDON, PRIMROSE HILL SHALL BE

IN CENTRE HOLD A BISHOP'S SEE

AROUND THE WORLD MEN'S THOUGHTS WILL FLY

QUICK AS THE TWINKLING OF AN EYE.

AND WATER SHALL GREAT WONDERS DO

HOW STRANGE. AND YET IT SHALL COME TRUE.

THROUGH TOWERING HILLS PROUD MEN SHALL RIDE

NO HORSE OR ASS MOVE BY HIS SIDE.

BENEATH THE WATER, MEN SHALL WALK

SHALL RIDE, SHALL SLEEP, SHALL EVEN TALK.

AND IN THE AIR MEN SHALL BE SEEN

IN WHITE AND BLACK AND EVEN GREEN

A GREAT MAN THEN, SHALL COME AND GO

FOR PROPHECY DECLARES IT SO.

IN WATER, IRON, THEN SHALL FLOAT

AS EASY AS A WOODEN BOAT

GOLD SHALL BE SEEN IN STREAM AND STONE

IN LAND THAT IS YET UNKNOWN.

AND ENGLAND SHALL ADMIT A JEW

YOU THINK THIS STRANGE, BUT IT IS TRUE

THE JEW THAT ONCE WAS HELD IN SCORN

SHALL OF A CHRISTIAN THEN BE BORN.

A HOUSE OF GLASS SHALL COME TO PASS

IN ENGLAND. BUT ALAS, ALAS

A WAR WILL FOLLOW WITH THE WORK

WHERE DWELLS THE PAGAN AND THE TURK

THESE STATES WILL LOCK IN FIERCEST STRIFE
AND SEEK TO TAKE EACH OTHERS LIFE.
WHEN NORTH SHALL THUS DIVIDE THE SOUTH
AND EAGLE BUILD IN LIONS MOUTH
THEN TAX AND BLOOD AND CRUEL WAR
SHALL COME TO EVERY HUMBLE DOOR.

THREE TIMES SHALL LOVELY SUNNY FRANCE
BE LED TO PLAY A BLOODY DANCE
BEFORE THE PEOPLE SHALL BE FREE
THREE TYRANT RULERS SHALL SHE SEE.

THREE RULERS IN SUCCESSION BE
EACH SPRINGS FROM DIFFERENT DYNASTY.
THEN WHEN THE FIERCEST STRIFE IS DONE
ENGLAND AND FRANCE SHALL BE AS ONE.

THE BRITISH OLIVE SHALL NEXT THEN TWINE

In marriage with a German vine.

Men walk beneath and over streams

Fulfilled shall be their wondrous dreams.

For in those wondrous far off days

The women shall adopt a craze

To dress like men, and trousers wear

And to cut off their locks of hair

They'll ride astride with brazen brow

As witches do on broomstick now.

And roaring monsters with man atop

Does seem to eat the verdant crop

And men shall fly as birds do now

And give away the horse and plough.

There'll be a sign for all to see

Be sure that it will certain be.

Then love shall die and marriage cease

And nations wane as babes decrease

AND WIVES SHALL FONDLE CATS AND DOGS

AND MEN LIVE MUCH THE SAME AS HOGS.

IN NINETEEN HUNDRED AND TWENTY SIX

BUILD HOUSES LIGHT OF STRAW AND STICKS.

FOR THEN SHALL MIGHTY WARS BE PLANNED

AND FIRE AND SWORD SHALL SWEEP THE LAND.

WHEN PICTURES SEEM ALIVE WITH MOVEMENTS FREE

WHEN BOATS LIKE FISHES SWIM BENEATH THE SEA,

WHEN MEN LIKE BIRDS SHALL SCOUR THE SKY

THEN HALF THE WORLD, DEEP DRENCHED IN BLOOD SHALL
DIE.

FOR THOSE WHO LIVE THE CENTURY THROUGH

IN FEAR AND TREMBLING THIS SHALL DO.

FLEE TO THE MOUNTAINS AND THE DENS

TO BOG AND FOREST AND WILD FENS.

FOR STORMS WILL RAGE AND OCEANS ROAR

When Gabriel stands on sea and shore

And as he blows his wondrous horn

Old worlds die and new be born.

A fiery dragon will cross the sky

Six times before this earth shall die

Mankind will tremble and frightened be

For the sixth heralds in this prophecy.

For seven days and seven nights

Man will watch this awesome sight.

The tides will rise beyond their ken

To bite away the shores and then

The mountains will begin to roar

And earthquakes split the plain to shore.

And flooding waters, rushing in

Will flood the lands with such a din

That mankind cowers in muddy fen

And snarls about his fellow men.

He bares his teeth and fights and kills

And secrets food in secret hills

And ugly in his fear, he lies

To kill marauders, thieves and spies.

Man flees in terror from the floods

And kills, and rapes and lies in blood

And spilling blood by mankind's hands

Will stain and bitter many lands

And when the dragon's tail is gone,

Man forgets, and smiles, and carries on

To apply himself - too late, too late

For mankind has earned deserved fate.

His masked smile - his false grandeur,

Will serve the Gods their anger stir.

And they will send the Dragon back

To light the sky - his tail will crack

Upon the earth and rend the earth

And man shall flee, King, Lord, and serf.

But slowly they are routed out

To seek diminishing water spout

And men will die of thirst before

The oceans rise to mount the shore.

And lands will crack and rend anew

You think it strange. It will come true.

And in some far off distant land

Some men - oh such a tiny band

Will have to leave their solid mount

And span the earth, those few to count,

Who survives this (unreadable) and then

Begin the human race again.

But not on land already there

But on ocean beds, stark, dry and bare

Not every soul on Earth will die

As the Dragon's tail goes sweeping by.

Not every land on earth will sink

But these will wallow in stench and stink

Of rotting bodies of beast and man

Of vegetation crisped on land.

But the land that rises from the sea

Will be dry and clean and soft and free

Of mankinds dirt and therefore be

The source of man's new dynasty.

And those that live will ever fear

The dragon's tail for many year

But time erases memory

You think it strange. But it will be.

And before the race is built anew

A silver serpent comes to view

And spew out men of like unknown

To mingle with the earth now grown

Cold from its heat and these men can

Enlighten the minds of future man.

To intermingle and show them how

To live and love and thus endow

The children with the second sight.

A natural thing so that they might

Grow graceful, humble and when they do

The Golden Age will start anew.

The dragon's tail is but a sign

For mankind's fall and man's decline.

And before this prophecy is done

I shall be burned at the stake, at one

My body singed and my soul set free

You think I utter blasphemy

You're wrong. These things have come to me

This prophecy will come to be.

On the outer wrappings of the supposed scrolls:

I KNOW I GO - I KNOW I'M FREE

I KNOW THAT THIS WILL COME TO BE.

SECRETED THIS - FOR THIS WILL BE

FOUND BY LATER DYNASTY

A DAIRY MAID, A BONNY LASS

SHALL KICK THIS STONE AS SHE DOES PASS

AND FIVE GENERATIONS SHE SHALL BREED

BEFORE ONE MALE CHILD DOES LEARN TO READ.

THIS IS THEN HELD YEAR BY YEAR

TILL AN IRON MONSTER TREMBLING FEAR

EATS PARCHMENT, WORDS AND QUILL AND INK

AND MANKIND IS GIVEN TIME TO THINK.

AND ONLY WHEN THIS COMES TO BE

WILL MANKIND READ THIS PROPHECY

BUT ONE MANS SWEETS ANOTHER'S BANE

SO I SHALL NOT HAVE BURNED IN VAIN.

And this one was "found in a scroll in another jar":

THE SIGNS WILL BE THERE FOR ALL TO READ

WHEN MAN SHALL DO MOST HEINOUS DEED

MAN WILL RUIN KINDER LIVES

BY TAKING THEM AS TO THEIR WIVES.

AND MURDER FOUL AND BRUTAL DEED

WHEN MAN WILL ONLY THINK OF GREED.

AND MAN SHALL WALK AS IF ASLEEP

HE DOES NOT LOOK - HE MAY NOT PEEP

AND IRON MEN THE TAIL SHALL DO

AND IRON CART AND CARRIAGE TOO.

THE KINGS SHALL FALSE PROMISE MAKE

AND TALK JUST FOR TALKINGS SAKE

AND NATIONS PLAN HORRIFIC WAR

The like as never seen before

And taxes rise and lively down

And nations wear perpetual frown.

Yet greater sign there be to see

As man nears latter century

Three sleeping mountains gather breath

And spew out mud, and ice and death.

And earthquakes swallow town and town,

In lands as yet to me unknown.

And christian one fights christian two

And nations sigh, yet nothing do

And yellow men great power gain

From mighty bear with whom they've lain.

These mighty tyrants will fail to do

They fail to split the world in two.

But from their acts a danger bred

An ague - leaving many dead.

AND PHYSICS FIND NO REMEDY

FOR THIS IS WORSE THAN LEPROSY.

OH MANY SIGNS FOR ALL TO SEE

THE TRUTH OF THIS TRUE PROPHECY.

CHAPTER 9

Mother Shipton is among the most mentioned of all prophets and prophetesses ever. She could be even called the British Nostradamus. Many 20th century books about prophecies mention her. Yet, at the same time, could everything concerning her be just a conglomeration of of legends and myths from old chapbooks and expanded and elaborated through centuries? Indeed, the birth of legends from chapbooks is happening even today in some parts of the globe and probably will continue to occur. In the northeast Brazil, for example, chapbooks attributing prophecies to a priest called Cicero and to a past messianic leader who both lived about 100 years ago are sold in some religious festivities in small towns, but nobody can be sure about what these prophets really predicted. In a few of these chapbooks, even a seven-meters-tall hairy leg announces the end of the world. Would you believe in the predictions of a tall talking hairy leg? I imagine you wouldn't, but some people do believe, as some peasants might, in werewolves, vampires, etc. Sometimes, some beliefs are part

81

of tradition and are linked to familiar teachings. Anyway, Mother Shipton prophecies and legends were born in a similar way, while other myths and legends are still coming to life through mankind's imagination. All of these kinds of topics are studied internationally at academic level today.

In the Middle Ages and in the immediate centuries following it, many times prophecies were attributed to religious men and women, priests, saints, and bishops and were considered authentic. Once a false prediction is printed as an authentic one, it is hard to stop the chain that will follow the seed. From time to time, a 20th century book printed in continental European countries such as Germany, France, Spain, Italy, revealed (or maybe created) a suspicious prophecy. This is done by attributing it to a person and saying it is authentic, sometimes without any kind of further information or research. In the case of Mother Shipton, the 17th century chapbooks helped to provide the foundation to construct her legendary life and fame, but other books printed in the following centuries, mainly in the last 150 years, provided a further touch; and perhaps other books will add a cent or two in the 21st century. In this chapter, we will see excerpts from some of these books – in English as well as foreign languages - to understand more easily how a legend can spread throughout the world and become, to some extent, part of our global history. A legend was born in the 17th century; now let's see how it grew and developed in the 20th century, turning it into what might be considered "history."

In *Book of Witches*, by Oliver M. Hueffer, published in 1908, the author presents the prophetess this way:

> Mother Shipton was a native of the gloomy forest of Knaresborough, in Yorkshire, a famous forcing-ground of Black Magic." And the author also believes that "no doubt Mother Shipton was as gifted as her fellows in the customary arts of witchcraft, but tradition draws a merciful veil over her exploits as poisoner or spell-caster. Her fame rests upon her prophecies, and wether she actually uttered them herself or they were attributed to her after the event, by admiring biographers, they have ever since been accepted as worthy of all respect even when not attended by the result she anticipated.

In *The Prophets and Our Times*, written by Rev. R. Gerald Culleton and originally published by the author in 1941 and 1943 in the US, a book with Imprimatur from the Catholic Church, and still in print today, we will find a Mother Shipton prophecy without further explanation about the prophetess, except the year she supposedly died (1551). The prophecy is the same printed in Chapter 3, but does not have the end-of-world prediction for 1881. Here is it how it was printed:

> Carriages without horses shall go,
> And accidents fill the world with woe.

Around the world thoughts shall fly
In the twinkling of an eye.
The world upside down shall be
And gold be found at the root of a tree.
Through hills man shall ride,
And no horse be at his side.
Under water men shall walk,
Shall ride, shall sleep, shall talk.
In the air men shall be seen,
In white, in black, in green;
Iron in the water shall float,
As easily as a wooden boat.
Gold shall be found and shown
In a land that's now not known.
Fire and water shall wonders do,
England shall at last admit a foe.

There is also this one:

All England's sons that plow the land
Shall oft be seen with book in hand.
The poor shall then most learning know,
And water wind where corn doth grow;
Great houses stand in farflung vale
All covered o'er with snow and hail.
Taxes for blood and war
Shall come to every door.
And state and state in fierce strife
Will seek after each other's life.
But when the North shall divide the South
An Eagle shall build in the Lion's mouth.
In London Primrose Hill shall be.
Its center hold a Bishop's See.
Three times shall lovely France
Be led to play a bloody dance;
Before the people shall be free,
Three tyrants rulers shall she see;
Three times the people's hope is gone,
Three rules, in succession, be ---
Each sprung from different dynasty.
Then when the fiercest fight is done,
England and France shall be as one.
The British olive next shall twine
In marriage with the German wine.

England shall at last admit a Jew,
The Jew that once was held in scorn
Shall of a Christian then be born,
A house of glass shall come to pass
In England – but alas, alas!
A war will follow with the work
Where dwells the Pagan and the Turk.

And the last one:

A great man shall come and go,
Men walk beneath and over streams.

Paula Clifford in her book *A Brief History of End-time* mentions an interesting prophecy attributed to a bridge in Knaresborough. The prophecy says the world will end when a bridge falls thrice and so far it has fallen twice. The author presents Mother Shipton legend as a historical fact, with her birth in 1488, her death in 1561, and some suspicious prophecies attributed to her.

The unabridged Hutchinson Encyclopedia has over 75,000 articles and is designed for homes and schools. This publication says Mother Shipton was an

English prophet. She acquired a large reputation in her native Yorkshire for correctly foretelling the future. She also made predictions, composed in verse, about succeeding centuries.

In *Psychics and Mediums in Canada*, published in 2005, Jean Porche and Deborah Vaughan state that

> At the beginning of the sixteenth century a British prophetess named Ursula Southeil, better known as Mother Shipton, was infamous for her accurate predictions. Born in a cave in the wilds of northern Yorkshire, Ursula demonstrated psychic tendencies even as a small child. In her heyday during the reigns of Henry VIII and Elizabeth I, her prophetic verses were very well known and feared throughout England. She foretold the invasion the invasion and defeat of the Spanish armada in 1588 and the Great Fire of London in 1666. She foresaw her own death by burning, which came to pass in 1561. Even today, her prophecies still prove uncannily accurate.

In the book *Las Grandes Profecías*, written by Roberto Buccellani and published in 1996 in Barcelona, Spain, through Editorial De Vecchi, an important European publishing house, the author compares Mother Shipton to Nostradamus and says that the following inscription is found in her tombstone, in towns called Clifton and Skipton:

Here ly's she who never ly'd
Whose skill often has been try'd
Her Prophecies shall still survive,
And she keep her name alive.

As I wrote earlier, this tombstone does not exist. The author further explains when and where she was born, when she died, etc., and presents her as a real prophetess.

In France, Gérard de Sède wrote *L'Étrange Univers des Prophètes*, published in 1977 by J'ai Lu. Gérard de Sède is better known for a book based on a manuscript written by Pierre Plantard, mentioning some allegations referring to the Priory of Sion deposited in the Bibliothèque Nationale de France. Mr. Sède published it as *Le trésor maudit de Rennes-le-Château*, which was used later as source material for the *Holy Blood, Holy Grail* as the authors clearly explain in the beginning of the book, which was used in turn by Dan Brown in *The Da Vinci Code*. In his book about prophecies, Sède strangely presents the 1881 prophecy using 1991, instead of 1881. The author also mentions the 1641 book, published initially in 1448 according to him (when Mother Shipton was not even born, according to tradition!). However, he explains that all myths have some truth in it. The author does not say clearly what is and what might not be myth concerning Shipton, he just presents the stories and prophecies, letting the task of sorting them out to the reader.

Also, Sylvie Simon in *Voyances Remarquables* makes curious statements about Mother Shipton. She dedicates a whole chapter to the seeress about "modern discoveries announced in the 15th century and printed in the 16th century," (which are most likely 19th century forgery). Even if, according to tradition, Mother Shipton had existed, and she was born in 1486-1488, there is no reference to her making predictions when she was 12 or 14. Sylvie Simon mentions a 15th century book called *Liber Elemonsinatoris*, which includes prophecies similar to the ones attributed to Shipton, but even if this is correct, there can be no relation between this book and our legendary prophetess (because of the dates). She also mentions the Kirby Cemetery prophecy.

Another book mentioning Mother Shipton's 1991 prophecy was published in Italy by Franco Cuomo in a book called "Le Grandi Profezie – una nuova chiave di lettura delle più celebri predizioni della storia". In this book the author says:

> The world should end categorically, according to her, in 1991. This is what one reads in her predictions, which circulated in her name in the first half of the 17th century, united in 1641 in a book...

As it is clear now, the supposed Mother Shipton prophecy about the end of the world in 1991 was a 20th century adaption to the 1881 prophecy. And this was never predicted by Mother Shipton, it was invented a few years before "her 19th century end" by an English writer. But Cuomo says clearly Mother Shipton was a real prophetess.

In contrast to the 20th century writers and publications who helped build the idea Mother Shipton really existed and was a real prophet,other books introduce her prophecies but also clearly explain that her life is more probably a legendary one.

In *Lexikon der Prophezeiungen* (published in 2000 and 2001 in Germany), written by Karl L. von Lichtenfels, an author I met in Austria a few years ago, he makes this statement about this seeress, which is partially accurate: "When she was 24, Ursula was well-known for her drawn-up forecasts of future events, often in mysteries and rhymes."

Then he mentions *Collections of Prophecies* written by William Lilly and published in 1646; he also says in 1641 there was a "rediscovery" of her prophecies. Finally he also mentions inventions and forgeries.

In Italy, *Le Profezie*, written by M. Centini and C. Bocca, and published in 1994, explains clearly about many legendary predictions attributed to Mother Shipton and their authenticity issue.

In *The Story of Prophecy in the Life of Mankind,"* written by Henry James Forman and copyrighted in 1936 in the US, the author refers to Mother Shipton as someone "whose name has become legendary in the English-speaking world. She is assumed to be the Progenitress of Punch." The author explains clearly that even her existence is questioned.

In the *Millennium Book of Prophecy,* published in New York in 1994, written by John Hogue, the author presents the following prophecy as authentically hers:

> There shall come the Son of Man,
> Having a fierce beast in his arms,
> Whose kingdom lies in the Land of the Moon,
> Which is dreadful throughout the whole world.

At the end of the book, the author says the following, referring to Mother Shipton:

> Any factual account of Ursula Southiel, the Yorkshire witch better known as 'Mother Shipton' suffers from a heaping helping of legend and wives' tales. The same is the case for her poetic augurings, which suffer from a witches' blend of forgeries from would-be prophets."

After the author's introduction, her life is presented, including how she was born and died, etc. The fact is that if the reader does not read the above at the end of the book in an appendix, but only her prophecies in the book's body, the reader will believe Mother Shipton was a real prophet.

The *Encyclopedia of Prophecy,* published in 1978 in New Jersey and written by Omar V. Garrison says the following about Mother Shipton:

Legendary English witch and soothsayer, known as the Yorkshire Sibyl, is supposed to have been born at Dropping Well, Knaresborough, Yorkshire, in about the year 1486. No biographical data concerning her is based upon trustworthy sources.

The *Encyclopedia of Claims, Frauds, and Hoaxes of the Occult and Supernatural* from the James Randi Educational Foundation presents elucidating information about Mother Shipton:

No reference to Mother Shipton prior to 1641 is in existence. It is thus difficult to determine whether this English prophet actually existed as she is represented in folklore, though writings seriously ascribed to her are being reproduced even today. There were several women who claimed to be her, but it is a Yorkshire claimant who has won the title.

Mother Shipton was Ursula Southill (or Sowthiel, or Southiel), the incredibly ugly daughter of Agatha Southill, known locally herself as a powerful witch. She is supposed to have been born in a cave at

Dropping Well, Knaresborough, Yorkshire, in 1488, and because of her unfortunate appearance and reputed powers, was widely rumored to be the child of Satan.

Sometime about 1512, she married a wealthy builder from York named Tobias Shipton. She soon attained considerable notoriety throughout England as "The Northern Prophetess," and her prognostications received great public attention, were printed in pamphlets, and were widely distributed. Though copies of these publications still exist, most of what can be found today are mere forgeries, and many meteorological and astrological almanacs published as late as the nineteenth century used Mother Shipton's name freely. An 1838 book gives an idea of the overblown claims made for such tomes. It is titled *The New Universal Dream-Book; or The Dreamer's Sure Guide to the Hidden Mysteries of Futurity —— By Mother Shipton.*

A 1686 book attributed to Edwin Pearson, *The Strange and Wonderful History of Mother Shipton,* because of its similarity to another book, *Life and Death of Mother Shipton,* was probably actually written by Richard Head, who also wrote *The English Rogue,* a racy account of his experiences with various tricksters, cheats, and rascals of his day.

The Mother Shipton entry ends this way:

Here ly's she who never ly'd
Whose skill often has been try'd

Her Prophecies shall still survive,
And she keep her name alive.

This is said to be the only such tribute to a witch in all of England, since the usual memorial——if there is any——consists of nothing more than a cairn of stones to mark the spot where such a person was hanged or burned.

The *Dictionary of National Biography* published by Oxford University Press, is another source revealing the other side, summarizing Mother Shipton as someone who might be nothing more than a myth:

Mother Shipton, reputed prophetess, is, in all likelihood a wholly mythical personage. No reference to her of earlier date than 1641 is extant. In that year there was published an anonymous tract entitled *The Prophecies of Mother Shipton in the Reigne of King Henry the 8th, foretelling the death of Cardinall Wolsey, the Lord Percy, and others, as also what should happen in insuing times.* According to this doubtful authority, Wolsey, after his nomination to the archbishopric of York, learnt that 'Mother Shipton' had prophesied he should never visit the city of York, and in consequence...

CHAPTER 10

I t's certain that some local businesses might have interest in building a the legend around Mother Shipton; for example, local tourism-related firms make money from Shipton's tale. Could money be the secret reason for some recent lies, while politics was the secret reason for the legend's growth in the 17th century?

One may say Mother Shipton had a good reason for being created in the 17th century, when her prophecies were reprinted and used for political use, such as in *Mercurius Propheticus or a Collection of Some Old Predictions*, a 12-page booklet published in 1643. Though her legend was initially manipulated with political interest in mind, in the following years, her fame would grow.

As Owen Davies says in *Witchcraft, Magic and Culture, 1736-1951*, published in 1999 by the Manchester University Press, over the time, the original prophecies published in the seventeenth century were also occasionally added to by chapbook publishers in order to update the relevancy of the content.

The author also says that

new prophecies attributed to Shipton were also generated in oral culture by communities seeking explanations for momentous local events. Thus when a railway viaduct collapsed during the construction of the line between Harrogate and York, the locals believed Mother Shipton had prophesied it would happen, though no such prophecy was heard of until after the disaster. In 1879 a rumour circulated in Somerset that Mother Shipton had predicted that the great Ham Hill stone quarry would be swallowed up by a tremendous earthquake on Good Friday of that year. Great alarm was felt in the neighborhood. Some left the area before Good Friday, while those that stayed removed all their crockery. When the dread day arrived, nothing untoward occurred and daily life resumed as normal. However this false alarm apparently did little to dispel the faith in Mother Shipton.

In a 1852 book, *Memoirs of Extraordinary Popular Delusions and the Madness of Crowds*, written by Charles Mackay, we have an idea about the faith on the prophetess in the 19th century in England:

The prophecies of Mother Shipton are still believed in many of the rural districts of England. In cottages and servants' hall her reputation is great; and she rules, the most popular of British prophets, among all the uneducated, or half-educated, portions of the community. She is generally supposed to have been born at Knaresborough, in the reign of Henry VII, and to have sold her soul to the Devil for the power of foretelling future events.

So politics and lack of education could be part of the reason for her legendary existence.

Concerning the further development of the legend, three different forces contributed to it in the 20th century. The first is related to money. Some firms might be directly involved in spreading the legend. Dropping Well Estate Ltd., for example, published a book called *The Life and Prophecies of Ursula Sontheil Better Known as Mother Shipton Carefully Compiled in the 1950s*. Maybe this was one of the first books announcing the end of the world for 1991, maybe the first one using the fake 1881 prophecy changed. Today it's not possible to know exactly when 1881 was turned into 1991 and who did it for the first time, but this probably happened in the middle of the 20th century. Naturally, some local firms are interested in keeping the legend, as it means business.

The second worthwhile fact related to the development of the legend is the globalization of publishing. If we check the most important books about prophecies printed in continental Europe until the end of the 19th century, in German, French,

Italian, etc. we will find no reference to Mother Shipton, except maybe in some rare occurrences. This leads to the conclusion that before the 1900s our legendary prophetess was not well-known outside of England or maybe outside of the English-speaking world, as she was rarely mentioned outside of her native land. It was only in the 20th century that many compilations of prophecies, from different countries, included her prophecies and biographical information. As the global publishing market grew, Mother Shipton became global. Some way, this is related to globalization and the growth of capitalism in the Western world after the Second World War.

The third aid Mother Shipton received in the 20th century came and still comes from the Internet (although there is no solid evidence that Ursula Southtell or Mother Shipton existed, the legendary aspects of her life grew and became known in many Western countries before the Internet age and continue to grow). While books show different sides of the story, truth, and lie, websites tend to introduce Mother Shipton as a real seeress and her prophecies as real, many times without their respective real history. These include those published in Australia (by Nexus). The Internet definitily has helped Mother Shipton to become really global and to spread all the folklore related to her all over our planet.

Now, did she ever exist? The gap between her supposed existence and the first book about her is long enough to allow inventive facts. At the same time, if there is no evidence Mother Shipton existed, there is also no evidence she didn't exist, and she might have lived as a peasant or any common woman.

In *Shell Games: Studies in Scams, Frauds and Deceits (1300-1650)*, edited by Margareth Reeves, et al., published in Toronto, at the Victoria University, and Centre for Reformation and Renaissance Studies in 2004, David A. Wilson points out,

> *The Prophesie of Mother Shipton* operates on the principle that no introductions are necessary and seems to assume that its audience must have been familiar with the prophetess.

Even though she may have had some local reputation at the time if she existed, she was not mentioned by John Leland on his 1540 tour, so she was probably not famous when alive, supposing she lived at that time.

The 1930 book called *Mystery and Romance of Astrology*, written by C. J. S. Thompson, analyzing the prophetess, he mentions that

> it is probable that the legend of Mother Shipton originated with some eccentric old woman who lived near Knaresborough and acquired a

local reputation for prognosticating, and telling fortunes, as it is started, perhaps with some truth.

Was she was really the fruit of inventive minds, and how does that relate to her existence and portraits (images) of her? Helen Ostovich, E. Sauer, and Melissa Smith state in *Reading Early Modern Women: An Anthology of Texts in Manuscript and Print, 1550-1700*, published in 2004,

the various surviving portraits of the secular prophet Mother Shipton of Knaresborough, Yorkshire, reveal the ambivalent attitudes toward women prophets in the sixteenth and seventeenth centuries.

Some pictures show her as a witch, others as an astrologer, and others as a respectable and well-dressed woman. Thus, how were these images of her created, in case she existed (if so, who was she?)

It's only possible to know that after she was born -- on printed paper in the 17th century or maybe from tradition or even from a real woman years before -- her legends spread during the following centuries in England, mainly in the North during the first centuries. Later she became known globally in the 20th century, particularly in the second half, once books and translations became common and accessible to people throughout the world.

100

The place where Shipton allegedly born is a beautiful place to visit. It even seems hard to imagine a wicked witch called "Devil's child" living in a such wonderful place. As we can't know into which direction a nocturnal moth known as Mother Shipton's Moth will fly, we can't be sure about what will happen to Shipton's future. Maybe she will become a goddess, maybe her legend will grow, maybe it will disappear. Maybe new prophecies will be attributed to her on Internet, referring to 2012. So in case you find someday a website saying Mother Shipton predicted the end of world for 2012, 2081, or another future year, beware of the falsity and scams in humankind's mentality and history and remember what you read here. And if you don't, never mind. Because using her broom, Mother Shipton may go to an untouchable and invisible area where the unpredictable is prognosticated and the impossible happens through spells, enchantment, or a never-found magic wand. So even if she predicted the end of world for a certain year, she may change her prediction, postpone the end, saving the world again and again.

The Strange and Wonderful HISTORY
OF

Mother Shipton,

Plainly setting forth
Her prodigious Birth, Life, Death, and
Burial.
With an exact Collection of all her famous

PROPHECYS

More Compleat than ever yet before
published. And large Explanations,
shewing how they have all along been
fulfilled to this very YEAR.

Licensed according to Order.

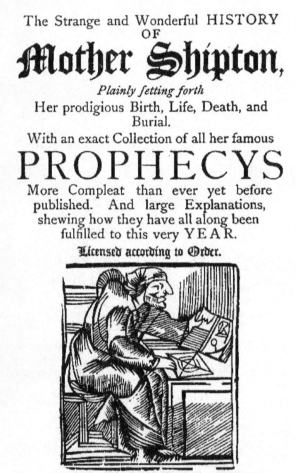

Printed for *W. H.* and sold by *J. Conyers* in *Fetterlane.*
1686.

The 1686 edition of The Strange and Wonderful History of Mother Shipton

DO YOU LIKE MYSTERIES OF HUMANKIND SUCH AS PROPHECIES OR LEGENDS?

One night, an Italian painter had a nightmare... He saw his city flooded looking from a certain room on the last floor of his house. The flood was huge, it couldn't be just a rain. Later he knew that at the same night, one of his sisters had exactly the same dream... and her dream had the same details of his dream... then he painted what he saw. He lives in Bologna, a city famous for its two towers. You can see them here and maybe the painting shows our post-global-warming time.

ONLY US$ 14.99

I n Selected Prophecies and Prophets, you will see

- How Nostradamus foresaw Putin as the Antichrist

- What might be the oldest prophecy in the world with the word "America." This prophecy has never been published before anywhere, and the author found it in a manuscript in a European library; it is translated from the Latin manuscript into English for the first time. The prophecy says the "Muslims will arrive in America." Could that be about the 9/11 2001, or is it about the future? Find it out.

- Many old prophecies translated from the original text from different languages (Old Italian, Old French, German, etc.) into English.

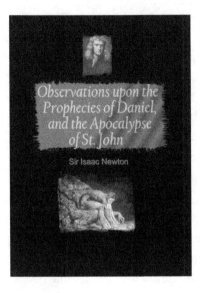

Originally published in 1733, written by Isaac Newton, *Observations upon the Prophecies of Daniel and the Apocalypse of St. John* is more than an interpretation of some prophecies contained in the Bible. It also reveals Newton's knowledge on many aspects regarding history, the primitive Church and the first followers of Christ. Still today, this is considered a great book.

ONLY US$ 9.50

"The Israelites in the days of the ancient Prophets, when the ten Tribes were led into captivity, expected a double return; and that at the first the Jews should build a new Temple inferior to Solomon's, until the time of that age should be fulfilled; and afterwards they should return from all places of their captivity, and build Jerusalem and the Temple gloriously, (...) which the Kings of the earth do bring their glory and honour. Now while such a return from captivity was the expectation of Israel, even before the times of Daniel, I know not why Daniel should omit it in his Prophecy. This part of the Prophecy being therefore not yet fulfilled..."

Know why the Russians are talking about this author and his theory...

The Antichrist:

A LEGEND? A MYTH? OR VLADIMIR PUTIN?

Published soon – **US$ 14.99**

- The Antichrist: 2000 Years of History

- August, 1999 and Putin in power

- August, 1999, the eclipse and the two crosses in the sky

- Nostradamus Prediction for the 1999 King

- Putin, the Richest Man and one of the most powerful men in the World?

- Putin President of the Red Russia and 666.666666...

- Are we close to the 3rd World War?

- The Global Economical Crisis before the 3rd War

- Secret Russian Weapons Predicted

You should be able to find these books in your local bookstore or at Amazon.com. Please email the author at admin@prophezeiung.com for more information about them and to buy any of them.

If you pre-order *The Antichrist* before it is published you will receive a 30% discount, and free shipping! Order any of *these books adverstised here*, and get a 30% discount. This offer may end at any moment. Free shipping is available in the US only.